W9-BIQ-357

SCARECROW!

Valerie Littlewood

DUTTON CHILDREN'S BOOKS · New York

CIP Data is available.

First published in the United States 1992 by Dutton Children's Books, a division of Penguin Books USA Inc., 375 Hudson Street, New York, New York 10014

Originally published in Great Britain 1992 by Julia MacRae, an imprint of the Random Century Group

Printed in Hong Kong FIRST AMERICAN EDITION
ISBN 0-525-44948-5
1 3 5 7 9 10 8 6 4 2

Acknowledgments

I would like to thank the following for their help and support in the preparation of this book: The staff of the M.A. Course in Narrative Illustration and Editorial Design at Brighton Polytechnic, particularly George, Chris, and John; John Key of Laughton for his farming advice; Norman Joplin for his help regarding model-farm scarecrows; Keith Stewart of Stewkie Ltd.; *The Farmer's Weekly*; Reading University Institute of Agricultural History; the relatives and friends who kept me well supplied with photographs, sketches, and descriptions of scarecrows from all over the country; all the farmers, gardeners, and enthusiasts whose creation and recording of scarecrows has made this book possible...and a special thanks to Alan, whose encouragement was constant and much appreciated.

Bibliography

Many books and magazines, old and new, were consulted while researching this book, but the main sources of information were: *The Scarecrow: Fact and Fiction*, by Peter Haining (Robert Hale, 1988); *The Scarecrow Book*, by James Giblin and Dale Ferguson (Crown Publishing, 1980); *British Toy Figures*, by Norman Joplin (Arms and Armour Press, 1987); *The Standard Cyclopedia of Modern Agriculture*, by Prof. Patrick Wright (Gresham, 1909); and *Farm Tools through the Ages*, by Michael Partridge (Osprey, 1973).

CONTENTS

The Scarecrow

Driving through the country, you see a farmhand working in the fields. But on the way back you notice the exact same figure in the exact same place! Instead of a real person, you probably are looking at a scarecrow.

Scarecrow, jack-of-straw, scarebird, tattybogle, or shoy-hoy — under many different names, these homemade figures have been used around the world to protect precious crops for over three thousand years. They are as old as the practice of farming itself.

We think of scarecrows as male or female dummies stuffed with straw. But over the centuries, farmers have invented many different "scarecrows" to protect their crops. Noisemakers, dead birds, smelly fires, pieces of cloth, shiny metal objects — even live people have been employed.

Since the time of ancient Egypt, paintings, plays, and poetry have featured them. Because of their frightening appearance, they often represent sinister spirits in myths and legends. But in children's storybooks, scarecrows have frequently been portrayed as loyal and friendly companions.

Beastly Birds

Starlings, ravens, pigeons, sparrows, geese . . . the list of feathered enemies goes on and on.

According to some farmers, the scarecrow is misnamed, since the crow is one bird they might not want to scare away. Feeding crows can control the populations of an insect called the corn borer and other worms and grubs. These small pests can destroy more crops than birds themselves do by eating seeds. However, many American farmers still want to rid their fields of crows, and in the late 1800s, town meetings were held where people argued for and against killing them.

Large numbers of birds can threaten a farmer's livelihood by destroying a whole crop. If the farmer grows food for subsistence, his or her entire family's survival may be at stake. Thus the task of scaring away birds has been taken very seriously in most cultures. The Creek Indians in what is now Alabama, Florida, and Georgia designated entire families to watch over the tribe's cornfields. In ancient Japan, the farmers ceremoniously burned their scarecrows, or *kakashi,* after each harvest as an offering to the spirit god of the fields as he departed for the year.

Clappers and Callers

Even more effective than a scarecrow that looks human is a human scarecrow. Early British records show that it was often the job of small boys or men too old for strenuous farm labor to go out into the fields and scare away the birds. Sometimes girls too would work as bird shooers.

In all weather, the bird scarers spent hours in the fields, throwing stones or flapping their arms while running and shouting. They also carried carved wooden clappers and rattles that made enough noise to frighten a whole flock of birds at one time.

Guarding a field or orchard all summer long was hard work. In bad weather, the human scarecrows had only crude huts made of mud and sticks to shelter them. To keep up their spirits, they would often sing songs and recite rhymes, or have contests to see who could hit the most birds with a slingshot.

When factories and mines opened up all over England in the early 1800s, the cost of labor rose and it became very difficult for landowners to afford human scarecrows. But even today, in India and some Middle Eastern countries, live bird scarers are still hard at work.

Men and Women of Straw

There are as many different scarecrows as there are farmers to make them. Each scarecrow is an individual, in dress, attitude, and effectiveness.

Some stand with arms and legs stiff and straight. Others flap loosely, appearing to stride across the fields. Gruesome scarecrows hang from makeshift gallows or poles and swing in the wind.

Faces may be happy, sad, sinister, or even blank. Heads can be made out of pumpkins or plastic jugs. Gloves make good scarecrow hands, and usually a strange combination of skirts, pants, blouses, and jackets covers the scarecrow's body. Sometimes scarecrows wear sunglasses or goggles. They also carry a great variety of noisemakers and fake weapons, including streamers, tin cans, flags, sticks, and trash-can lids.

Field Duty

Through rain, wind, snow, sleet, and blazing sun, scarecrows stay at their post, guarding crops from sowing to harvest. In the spring they may help protect young corn; in the summer, peas and cabbages; in the autumn and winter, the early sowings of wheat.

The traditional stuffed scarecrow is the most common, but now there are also inflatable plastic scarecrows and even a self-inflating scarecrow that leaps up, shrieks loudly, and glows in the dark.

Some farmers fly flags with huge eyes painted on them or play tapes of a loud humming sound over a speaker system. In England, a device called an automatic crop protector was invented by a fireworks company. A metal box with three arms sat on top of a pole. Every forty-five minutes, a cap of gunpowder exploded, pulling on a rope, which in turn made the metal arms flap up and down and caused a great racket.

According to many farmers, the best scaring device of all is a huge hawk-shaped kite that moves across the field on wires, casting the ominous shadow of a bird of prey.

Stars and Stripes and Streamers

Many towns and cities have community gardens, where an area of land is subdivided into plots and allotted to different tenants. When food was scarce during World War II and in the Great Depression, this type of garden played a vital role in contributing to the country's food supply. Each family was assigned a plot of land and grew much of their food there. In America, scarecrows made a comeback during this period, both in the cities and on the farms.

Today there are no Victory Gardens the way there were during wartime, but community gardening groups often take over abandoned lots in order to raise fresh vegetables and add greenery to the city landscape.

At the height of the growing season, these urban gardens are protected by an array of scaring devices. The traditional scarecrow is helped by streamers, flags, tin cans on strings, and old wheel hubs—anything that flutters, sparkles, or rattles. Sometimes a dead bird is even hung from a stick as a warning, a method borrowed from the Zuni Indians.

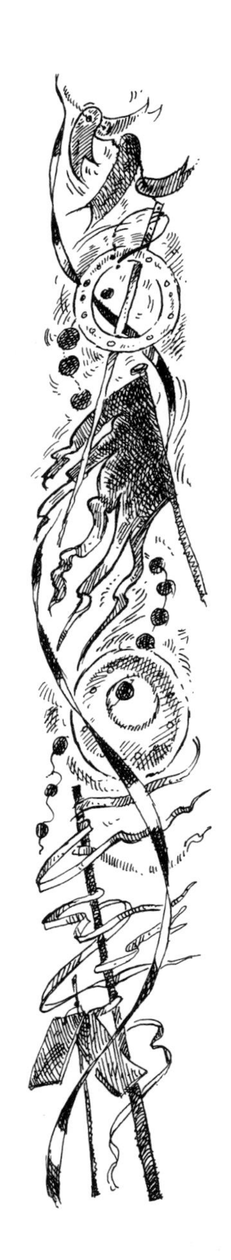

Garden Ghosts

Home vegetable gardens are particularly vulnerable to pests because of their small size. Scarecrows, sometimes lovingly crafted by the whole family, may be present to help protect the rows of radishes, tomatoes, beans, and lettuces.

Private gardeners are often ingenious in the details of their scaring devices. Some hang long strips of plastic from sticks and string cotton over new shoots. To scare away moles, a bottle stuck in the ground with its neck poking up works well. Whenever the wind blows, the bottle makes an eerie wailing sound. Slugs can be lured under leaves with bits of fruit and then removed.

At night, mirrors and tin lids that glint and shine in the light from street lamps and windows help frighten pests. Mice may be scared away by a fake cat head with sparkling marble eyes, hanging from a post. Gardeners also can buy inflatable plastic owls to peer down from the trees and keep away small birds.

Fanciful Farmyard Fun

Thanks to the most famous scarecrow of all, the scarecrow without a brain in *The Wizard of Oz,* by L. Frank Baum, people have an insight into a scarecrow's view of life. This much-loved character is based on a real scarecrow the author saw near his home. In England, Barbara Euphan Todd wrote about a scarecrow named Worzel Gummidge from Scatterbrook Farm, who became just as popular with England's children as Baum's figure did in America. The success of these characters in books, radio, television, and film has earned the scarecrow a lasting part in the world of make-believe.

Scarecrow costumes are always popular at Halloween. A scarecrow outfit can easily be made from an old hat, overalls, a jacket stuffed with straw, and maybe a carrot-shaped nose—all worn in tribute to the fictional heroes.

There have been many scarecrow puppets and dolls, but the first toy scarecrows probably belonged to model-farm sets. Just after World War I, manufacturers noticed a decline in the sale of toy soldiers and began to create tiny farm figures out of lead. Now these toy scarecrows for farm sets are made of nontoxic plastic.

Myths and Magic

Scarecrows are popular figures in legends and horror stories. In some of these spooky tales, spies, lookouts, and murderers disguise themselves as scarecrows. Dead bodies have reportedly been hidden inside them, and scarecrows have also hosted ghosts and spirits. At magic festivals, such as Halloween, they have been rumored to come alive, spiriting children away and dragging them down into the soil.

The Greek scarecrow figure has its origins in the myth of the god Priapus, the son of Dionysus and Aphrodite. Priapus was adopted by vineyard keepers when they saw that the birds feasting on their grapes were frightened away by his unusually ugly face and body. When the vineyard keepers spread the news about their unwitting scarecrow, other farmers carved wooden statues resembling Priapus and stood them in their fields to try to achieve the same effect. Since then Priapus has become the god of gardens, and he is often depicted with a club for protection and a sickle to encourage a good harvest.

Strange Sightings

Some thrifty farmers construct scarecrows from very unusual materials. Giant corrugated-iron ones with spinners instead of heads sometimes protect wheat fields. Sheets of plastic can be twisted into a bodylike shape; even an old suit of armor was used by a farmer in England in the early 1900s.

In addition to strange-looking scarecrows, you can also spot scarecrows in very strange places. On Guy Fawkes Day, a British holiday similar to the American Fourth of July, many old scarecrows come to the end of their lives on top of bonfires as part of the celebrations. And according to a very old German story, the citizens of one small kingdom actually protected themselves in battle by gathering all the wooden scarecrows in the area and propping them up on the battlefield. Thinking they were outnumbered, the enemy troops fled.

Making a Scarecrow

These general directions tell you how to make a small scarecrow. Please ask an adult to help you.

Take two strong sticks, one about six feet long and one about two feet long, and nail the short one to the long one to form a cross.

Stuff an old skirt and blouse or a pair of jeans or overalls with straw and tie them to the sticks with twine or rope. For the head, you can use almost anything: a straw-stuffed pillowcase, a plastic jug, a pumpkin. Draw a face on it with markers or add a mask.

Push the head down onto the top of the cross and tie it firmly in place.

Now that you have your basic scarecrow, you can add any details you like: a jacket, hat, gloves, boots.

For motion, add a plastic streamer, a scarf, a flag, or a stick. Create noise by hanging a tin can with pebbles in it from the scarecrow's hand.

Position your scarecrow so that he or she appears to stride out across the field or garden and plant the long pole firmly into the ground.

Stand back and see how many friends wave to the scarecrow as they go by, thinking it is you—and how few birds pay any attention at all!

2
STRONG STICKS
HAT
SCARF
GLOVES
PLYWOOD FOR DOG
PLASTIC FOR FLAG
STRAW FOR STUFFING

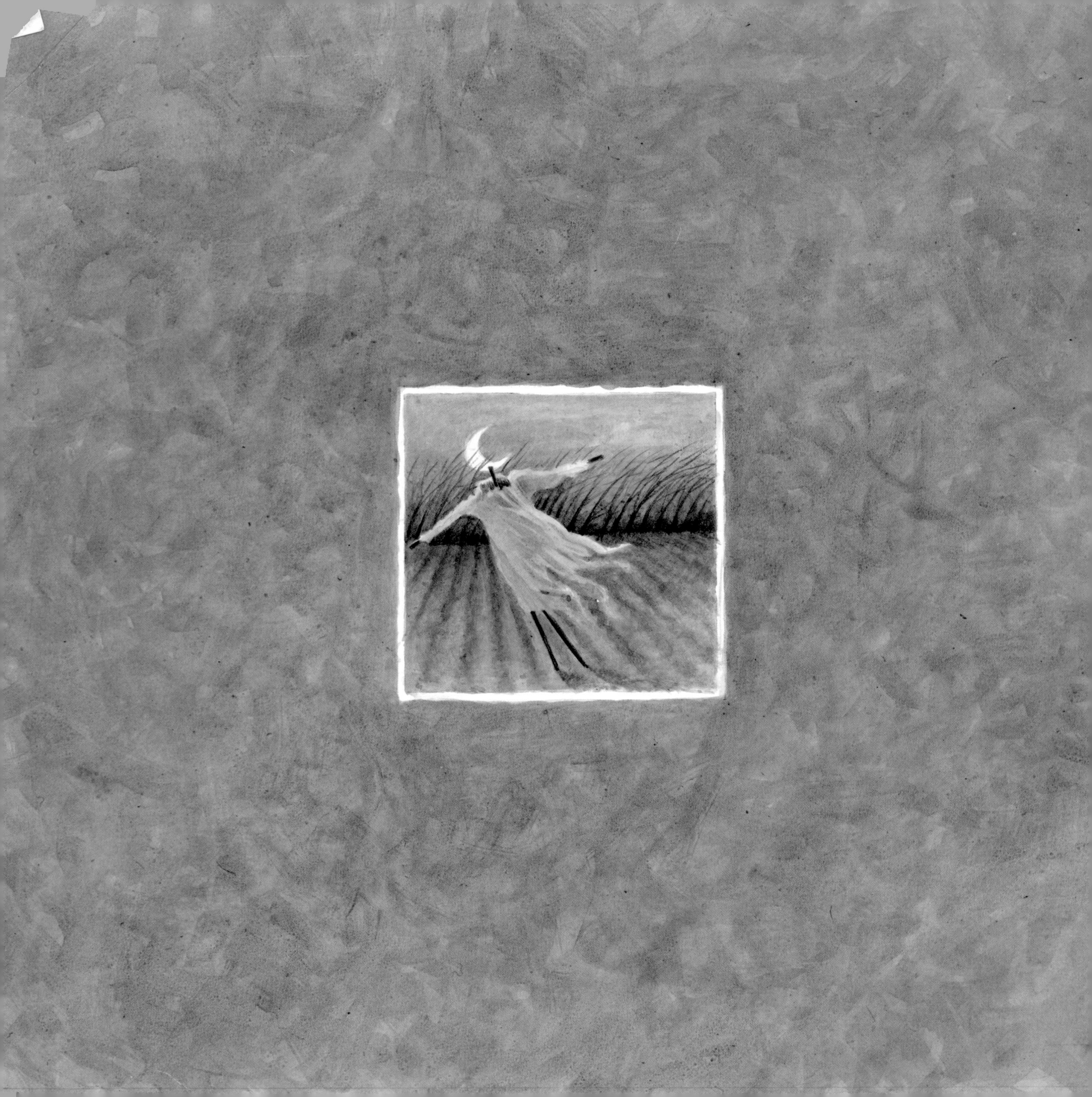